What you need to know <u>before</u> you cut your cable

or

How I learned to cut out the cable and still keep my sanity….

Roku and DIRECTV NOW Edition

By R.T. Mayhue

Copyright © 2019 R. T. Mayhue
All rights reserved.

Acknowledgements

One of the hardest things you'll ever do is to sit down every day and try to come up with something worthwhile to write about. But, when I finally got around to deciding it was time to get rid of my high priced cable bill, I realized there were a lot of other people in the same boat and someone needed to let them know what to expect and how they could go about doing it too. Which I soon found out is something the 'experts' seemed to be leaving out. After all, the whole thing sounds pretty simple, cut the cable, do 'x' and everything will be okay. We all know better.

Everyone has different circumstances and that's why one of the first things I'm going to tackle is the 'deciding' part of this equation so you can figure out which type of cord cutting is best for you, or even if the whole thing is flat out not for you. I also want everything to stand alone so that you can get the information you need to get up and running with the least pain possible. That's why I'll try to touch on as many specific combinations of services and solutions as I can even though the focus this book is about setting up DIRECTV NOW on a Roku.

I'm able to cover some of this because my brother, Rob, had already gone, full blown, cable free, 'commando' a while back. Shutting it all down, throwing up an antenna and coming up with a pretty cool setup using a box that could not only record the 'Over The Air' broadcast programs, it could even stream them to his TV and other devices around the house. Bear in mind, as a web host, website developer and programmer, my brother is a bit higher up on the 'geek' ladder than I am. In fact, he's my sounding board for making sure some of the 'boiled down' explanations I come up with in this book are reasonably accurate. Trust me, I'm not going to get too nitpicky over my explanations and don't really care if someone else wants to throw a flag. Because the point here is to pass along the information you need to get up and running, not bury you up to your neck in perfect precise explanations you couldn't care less about. After all, just because you can't work on the motor, doesn't mean you don't know how to drive a car.

Trust me, when it comes to technical stuff, your average geek doesn't have the slightest problem swerving off the road and getting deep into the weeds with some spectacularly yawn inducing stuff that's sure to make your eyes glaze over while they try to impress you with their smarts. I'll do everything I can to avoid that. Trust me, I'm not that smart, just ask my wife....

Speaking of which, this wouldn't be possible without her. She's not only my better half, she's my editor and, if necessary, my illustrator. *(Assuming I can pry her away from all her other artistic endeavors.)*

In any case, many thanks to Terri my wife and Rob, my brother.

Table of Contents

Acknowledgements	1
Why did I decide to do this?	10
See if any of this sounds familiar:	11
The Starting Point	15
Commando Mode	16
Streaming Meemies...	20
Roku	21
Streaming	26
WiFi - It's all about the speed	29
How to setup and use a Roku.	31
Which leads to a word about HDMI cables.	33
Now, back to setting up the Roku...	36
Adding DIRECTV NOW	41
Setting up the DIRECTV NOW service	43
Getting Started With DIRECTV NOW Streaming	46
Guide	46
Main Menu	47
Overlay Screen	50
Channel View	51

The Online DVR   **54**
    How to Record a show using the DVR   56
    Record a show from the Guide   57
    How to cancel the recording of a show or a series using the Guide   58
    Recording a program while it is in progress   59
    How to play recorded program   60
    Viewing a recorded program   61
    Bookmarks   62
    How to work with your Bookmarks   64
    The second way to turn Closed Captions off and on   65

Miscellaneous Stuff   **67**
    What if something isn't working right?   67
    Take it with you   69
    Headphones   70
    The single most optional accessory   71
    DIRECTV NOW Channel Timeout   73
    Setting up the Automatic HDMI port for the Roku   74
    The internet download speed indicator   75

Roku Mobile App   **76**

DIRECTV NOW online   **80**

Now what?   **82**

Looking to the future   **87**

Epilogue   **90**

What you need to know <u>before</u> you cut your cable

or

How I learned to cut out the cable and still keep my sanity….

Roku and DIRECTV NOW Edition

There are a lot of helpful places, books, websites, you name it out there that are great at recommending that you 'cut the cable' and quit wasting your money. But, in my experience, I couldn't find anyone to help with the part that comes after your cable is cut. You know the, 'What do I do now?' part.

Despite all that, I dove in anyway and this book is all about how I went about cutting my own cable in order to help someone else who may be thinking about doing the same thing.

Just for the record, this book isn't entirely about completely cutting the cable, lock, stock and COAX. This isn't about just using an antenna even though I'll cover that. This is mostly aimed right at those of us who don't really want to give up all the perks of cable, *(We're willing to make some sacrifices, but we all have a few favorite 'cable only' channels we can't seem to do without.)* it's just that we're really tired of paying through the nose for them. I'm going to go over, how I did what I did and why. I'll also do my best to explain how it all works so anyone interested in doing it will benefit. What I want to get out of the way, right off the bat, is that this can save you money if you do it right. But, like anything else, because your particular circumstances are probably going to be a little different, YMMV. *(Your Mileage May Vary.)*

Here's a little background. I'm a computer guy. Please don't hold that against me. But, before you throw your hands up and think I'm going to bombard you with computer type solutions and jargon, let me assure you, I'm not one of those guys. You know the ones, they work for a large box store and charge a fortune just to get them to show up at your door. Let alone how much extra you have to pay for the gimmicky car, white shirt and tie.

    I'm more of a results oriented computer fixer. If I can't fix it, I don't charge for it. And, I don't charge for looking. Granted, it didn't make me rich, but I've never liked the idea of profiting from someone else's pain. And, even though it doesn't seem like it, a screwed up computer can cause a lot of pain.

    Okay, enough of that, let's get started.

Why did I decide to do this?

One of the things that was driving me crazy, *(And probably you too, or you wouldn't be reading this.)* was just how much my cable bill was costing me. Especially, since in order to get the channels I really wanted, I had to get a whole lot of channels I wasn't even remotely interested in.

It's called 'bundling' and it's a result of the deals the cable providers have to make in order to get the channels everyone really wants to see. Quick show of hands, 'How many of you are watching the 'Baby' channel?' Yet, there it is in the bundle. In order for your cable company to make it appear that you're getting a lot of stuff in your particular 'bundle', they have to pad it with all those oddball channels they were forced to take in the deal they had to make with the content providers. ...And you're going to pay for it. That's why they sell it as 'Over 100 HD channels!' Concentrating on the number of channels in the bundle instead of focusing on the few you're really interested in.

Now, in their defense, the cable provider's point of view is that if they don't throw in all those other less popular channels in their bundles, those channels will eventually cease to exist without our forced support. Which makes me wonder what the advertising on those channels is supposed to be there for if not to pay the bills. The reality, from my standpoint, is it's more likely that early on they needed 'filler' channels to market the '100's of channels' angle. Now, they're kind of stuck with it.

See if any of this sounds familiar:

When we moved into our first house many, many moons ago, all we had was the local cable provider and a dream of being able to watch more than 5 channels. Believe it or not, one of the biggest draws turned out to be a slow moving radar image from the fine folks at NOAA showing the current cloud cover. In our defense, we live in Florida and for people who see almost the same weather forecast, day in and day out, it's amazing how weather obsessed we are. Now, this wasn't the kind of weather radar you're used to seeing today, we're talking about one of those phosphorescent green circular screens with an actual glowing arm sweeping around in a circle just like you see in old war movies. This was complete with occasional static riddled announcements read by a gravelly voiced gentleman who sounded like he clearly had smoked way too many unfiltered Camels. Cable back then only meant that you didn't have to struggle with an antenna to get all the mostly local channels and during 'Free Preview' weekends, you could even watch some HBO.

      Years went by and our cable provider was bought out by another company. The new guys definitely weren't known for their stellar support and after a few painful experiences, it was time for something else. That was when a flyer from DirecTV landed in our mailbox and they happened to be running a special that included a free year of the NFL Sunday Ticket. Of course, we jumped at the chance to be able to watch our hometown Steelers play again. *(Yes, I'm one of those.)* After the DirecTV installer hung the satellite dish off the side of the house, finished with the rest of his handiwork and we fired it up for the first time, I thought, for sure, I had also acquired brand new TV sets in the deal. The picture was just that good. Surprisingly, in all the time we had a DirecTV satellite dish, we only had a handful of brief signal outages due to weather so I can't really complain.

      But, then came FIOS. When the Verizon guy came around the neighborhood checking to see if I somehow might be interested in ditching my old DSL internet connection for a fiber optic connection that was roughly 10 times faster and they'd throw in cable TV to boot, I was all in. Truthfully, once again, it was like I had somehow managed to get, even better, brand new TVs. Of course, it helped that I'd already been through a couple of other FIOS installs with a few of my clients and I knew to totally re-cable my place ahead of time. Why did I bother with that? Because my house still had the original old COAX cable that the builder installed 15 years before and it definitely wasn't going to be up to snuff to handle the newfangled HD signals we were about to experience. Add to that, at the time, a typical FIOS installation was taking the better part of a 10 to 12 hour day and, since I work out of my house, I couldn't afford to be out of business that long. My FIOS installation only ran 4 hours start to finish. *(I think the installer was happier than I was.)*

After years of absolute rock solid uptime from Verizon, they sold the whole thing off and Frontier came to town. FIOS became their baby officially, and appropriately, on an April Fools Day. With outages, billing issues and businesses being down for days, almost every FIOS customer I knew were either affected somehow or lost something in the transition. During the process, I lost my old Verizon 'bundle', had to deal with three different levels of tech support to get my channels back up, and my bill jumped up about thirty bucks higher because Frontier couldn't match the 'bundle' I originally had. If that weren't bad enough, it seemed like every time I turned around my bill kept creeping up higher and higher.

I knew a long time ago that it was time to call it quits but, where do you even start?

The Starting Point

Anyone who's even remotely considering cutting the proverbial cable has to realize the single most important thing in the process, namely your 'significant other' or, if that doesn't happen to be you, then the person that this really matters to the most. Every house has one, we know who they are and they absolutely, positively have to be on board. This is especially true if that person also happens to be the one that pilots the cable remote control like Maverick handles an F-14A Tomcat. They don't just flip through the channels, ...they navigate. The TV doesn't just turn on when they push the power button, it winces. You get my meaning. Anyone who has that kind of dedication to a hand held device has to be handled delicately because, no matter how you look at it, this will be a life changing event for them.

Best course of action, don't jump in with both feet. You already have cable. Keep it around for a while as a safety net until you're sure you can get along with the new set up. When you're past all the setup hurdles, then you can eliminate the cable and just stick with the new setup.

However, there are also few things you need to know, no matter which route you take.

Commando Mode

First off, let's say your goal is to ditch your cable company completely. But, you still want be able to keep tabs on all the shenanigans happening on 'Wheel of Fortune'. Total cable cutting means you'll need some sort of an antenna to receive your local, over the air, broadcast TV signals. Before you fully commit yourself to being footloose and cable free, you'll also want to have a pretty good idea of which stations you can expect to be able to pull in at your humble abode. So, to help with that, jump on your computer and fire up your internet browser of choice and head on over to this website, https://www.fcc.gov/media/engineering/dtvmaps. This is the link to the Federal Communications Commission's official 'DTV Reception Maps' website and simply entering your zip code or address will produce a list of broadcast stations in your area with a map. If you click on any of the stations on the list they'll pop up on the map with 'tower' icons to show you where the station is broadcasting from in relation to you. There is also a surprising amount of detail, like signal strength and distance, for all the broadcast TV stations in your area to help you see what to expect before you decide to go 'Commando'. There are other websites out there as well and it doesn't hurt to get a second opinion, but most of them derive their information from the data the FCC has gathered at this site. Remember, the more information you have ahead of time, the better informed your choice becomes.

In case you've been living under a rock, a lot has changed since the days when we all used to have a pair of 'rabbit ears' hanging off our TV sets to drag in all those fuzzy, faded, static riddled, snowy images we used to think were so great. By the way, if you know what 'rabbit ears' are, or still recall that yours were stylishly decked out with generous clumps of aluminum foil, it's also is a sign of your age…

Back in 2009, the government mandated the end of analog broadcast signals so nowadays, the 'D' in 'DTV' stands for 'digital'. And, if you know anything about digital, you know we're basically talking about 'ones' and 'zeros' which, for our purposes, means you can simply substitute 'yes' or 'no'. That means, unlike the old analog broadcasts, when it comes to broadcast digital TV signals, they usually either work or they don't. Which generally tends to make reception an all or nothing proposition. In other words, with digital, you won't see those faded out 'ghosts' or doubled images and snow on your screen anymore. Instead, signal reception may come and go with a distant station or weak signal but, it won't just fade out or make everything look like an episode of 'Outer Limits'.

Unless the TV you're using is so old it comes complete with stone age cave drawings and a candle in the rear to light up the screen, it likely already has a digital tuner built right in. In that case, if you live in an area where the broadcast signal strength is considered 'Strong' or 'Moderate' you can get your hands on one of the new flat panel indoor antennas and try hooking it up to your TV. I'm not going to recommend one in particular because they're a moving target with new ones coming out all the time. Do your research and try to buy one that claims to be able to receive signals from the distance that the reception map on the FCC site shows you. If, for instance, your local TV stations are 25 miles away according to the reception map, get an antenna that can haul in signals from 35 miles away or better. If you're dealing with really distant signals or bad local terrain, you'll likely need an outdoor antenna.

Once you connect your antenna to your TV, you'll need to switch your 'input' or 'source' on your TV to 'Antenna' and have your TV run a scan to see what channels it can find. Once you have a baseline of local stations programmed into your TV you'll be able to make an informed decision as to whether or not this is the direction you want to go. But, you don't have to stop there. There are add-on devices that can even record the live TV broadcasts for you and even stream it to other TV's located throughout your house.

There are many broadcast TV DVR devices on the market today. For instance, there's the original DVR - **Tivo**, and then the **Tablo** which is a great choice for homes with multiple TVs. If you're familiar with Plex, HDHomerun is a great choice and recently, even Amazon got into the act with their Fire TV Recast. With devices like these, you can record the over the air broadcasts and stream directly to or playback the programs on other TV's via your home's WiFi network or, in some cases even stream your local programs over the internet to another location.

Generally speaking, to use one of these OTA (**O**ver **T**he **A**ir) DVR boxes, you connect your antenna to the box and run an HDMI cable from the box to your TV. All the tuning, recording and programming is handled by the box instead of your TV. Once it's set up and running, it will act a lot like your current cable set top box. To get the programs to show up on your home's other TVs, you can add streaming devices to them and download an 'app' from the manufacturer which will connect the TV to the OTA DVR, just like you would use it for 'Streaming' which we will cover, next…

Streaming Meemies…

If you decide going 'Commando' isn't your cup of tea and you miss some of the cable based programs you know and love, then it's time to check out the 'streaming' route. It's good to know that streaming services work a little differently than cable. You see, unlike cable, they have no equipment or installation involved and that allows all the streaming services to have a 'trial period' so you can run a shake down cruise and see if their offering fits the bill. If you don't particularly like the one you've chosen, you can always try another one. There are quite a few to choose from with more coming along all the time and none of them lock you into any sort of a long term commitment. That means, you can change things up any time you like.

But, before can even we get that far, you need something that takes the place of your old cable box to get the job done. Something that can take the stuff on the internet and make it show up on your TV. Namely, a streaming device. In this case, that's where something like the Roku comes in.

Roku

The first thing to clear up is, even though it may sound like it, a Roku is not a character in an old samurai movie. I bought my first Roku years ago. It was a Roku 3. Basically, we're talking about a little black trapezoidal shaped box only slightly bigger than a pack of cigarettes. *(Note: I know it's not 'PC' and who knows what the future will hold, but for now, the number of 'packs of cigarettes' has always proven to be a reliable unit of size measurement that's universally understood. Even by those of us who don't smoke.)* Compared to today's models, my Roku 3 was a fairly basic unit. But, it easily got the job done. In fact, when I decided to get into all this, I actually had to perform an archeological dig to unearth my trusty little Roku from behind a heavy dresser in my bedroom where it had fallen way back when. Which gives me an opportunity to point out that the remote on my Roku doesn't communicate like the typical remote on your TV or cable box does using IR. (infrared)

> *Editor: I told him he needs to explain 'infrared' a little better. Seriously, I had no idea what he was talking about. If you do, you can skip this little box and keep on going…*
>
> *At the risk of turning this into a high school science class, I'll do my best to explain infrared without making it sound like a Star Trek episode. Basically, infrared is a class of light waves that are just outside the visible light spectrum. Since you're dealing with light, even though it's lightwaves we can't see, think of your TV remote like a flashlight and you'll understand why you have to point the remote directly at your TV to get it to work.*

Instead my Roku remote uses RC, *(Radio Control or radio waves.)* which means, you don't need to have a 'line of sight' connection because you don't have to point it at the TV or cable box. I could even use the remote to run my Roku from another room if it wasn't too far away. All of which is very cool. That's why, even when mine fell behind my dresser, I didn't even notice because it still kept working just fine.

I originally bought it because my brother was developing websites for clients that he was setting up with their very own private Roku channels they could use to broadcast their 'In House' video content to their members directly to their TV sets. *(Which is great stuff for organizations and churches.)* As it turns out, he needed a guinea pig for testing, and I was it. Since then, I've used my Roku with Netflix, Amazon Prime, Plex and a raft of other Roku channels. While it wasn't for streaming 'Live TV' because there weren't any 'live' services at the time, it helped get my feet wet and learn a lot about how they work. Nowadays there are TVs on the market with Roku built right in, but they came along years after I bought my first Roku.

Fast forward to today and because I already had one, it made sense to use my Roku for my new streaming set up, not only because I still think it's the best bang for the buck, but because, at the same time I upgraded the COAX cable in my house to prepare for FIOS, I also went 'all in' and added a wired *(ethernet)* network. This means there is also a wired ethernet connection everywhere all my TVs are located. That way I could take advantage of the fact that my particular model of Roku 3 at the time *(4200R.)* not only had WiFi, it also had an ethernet plug-in for my network cable. So it was a 'win, win' for me.

> *Editor: Okay, sorry. Me again. I think he needs to explain what 'ethernet' is. Sounds vaguely like something dentists used to use to knock you out.*
>
> *First thing I'd like to say is, let's be glad they settled on calling it 'ethernet' because it's inventor wanted to name it after 'luminiferous aether' a term I won't explain but was shortened to 'ether' hence 'ethernet'. In any case, today's ethernet cable uses connectors that look a lot like giant phone plugs to connect computers and devices together or connect to the internet.*

Don't get me wrong, by no means is Roku the only game in town for streaming devices. There are several brands of streaming devices around, including (*In alphabetical order.*) Amazon Fire TV, Apple TV, Google Chromecast, NVidia Shield, which also doubles as a gaming console, speaking of which, Playstation's have Playstation Vue and even XBoxes can stream 'Live TV' services with a promise of DirecTV Now support in the future. Newer generations of Roku, Amazon Fire TV and Google Chromecast streaming 'sticks' plug directly into one of the HDMI plugs on your TV and use your home's WiFi for the internet streaming access.

My newest Roku Ultra, not only streams, but unlike my original Roku 3 4200R model, the Ultra's remote also controls my TV. It's 'enhanced' remote has an On/Off button and a volume control that operates the TV as well as a voice control button and the standard headphone jack on it for private listening. There's even a remote control 'finder' button located on the main box that's extremely handy since I have both a wife, *(You may remember her as call sign 'Maverick'.)* and a dog, who like to play 'keep away' with the remote. It's one of those things, you think you'll never need, ...until you need it.

Streaming

Whatever streaming device you settle on will most likely mean you'll be using WiFi to connect to your internet. This means you'll need to know just how good the WiFi signal happens to be where your TV is located. Bad signal = Bad experience. So, to avoid any heartache later, you can check the signal strength beforehand using a laptop or a cell phone. But, whatever streaming device you settle on will also be able to let you know how good the signal you're getting happens to be when you are setting up the connection for the first time.

Basically, you need to think of a streaming device as a mini computer. So it will require the stuff a computer needs, like a network connection. *(WiFi or wired.)* Power, which it will get from either a USB port on your TV or a power adapter that plugs into an electrical wall outlet. An input device, *(The remote.)* and an output device. *(The TV.)*

Next, if you've never used one, you really need to know how a streaming device actually works.

Streaming video is basically a series of files that are being downloaded by your particular streaming device. It would be nice if it were really that simple but, for our purposes, it's probably best to stick to the basics. So the process looks like this, the source, *(For example, Netflix.)* takes the raw video, packages it up in the most efficient way possible using various methods to compress it to the smallest size it can while breaking it up into a continuous series of easily digestible individual files. Each file becomes one of many thousands that are all being sent streaming down the internet to your device. As everyone knows, the internet is a wild and wooly place and sometimes the files arrive out of order or maybe one of them even gets mangled along the way and needs to be re-sent a second time. Your streaming device, in this case, the Roku, solves this problem simply by not sending anything on to your TV the very second it receives it. Instead, it reassembles and decompresses the files and then slightly delays the playback by storing a certain amount of the received video in a small storage area called a buffer. Depending on the kind of file being streamed, this buffer can contain minutes or seconds of video waiting to be sent on to your TV. This delaying technique, known as 'buffering', assures that your video can play uninterrupted even though it's really arriving a little bit late to the table. This effect is easy to see it you are watching a live stream that happens to be displaying a clock on the screen that you can compare to the real time at your location.

Your cable TV box however, follows a different set of rules. The cable contains all the signals, from all the available channels, all the time. Your old cable set top box merely has to change to the channel you are going to watch. This process happens, usually without delay. Streaming devices, on the other hand, only receive the 'channel' you've chosen to watch and the buffering process I described above means that, when you change from one channel to the next, there will be a slight delay while the Roku or whatever device you happen to be using, downloads, then begins buffering the channel, and finally starts sending the stream to the TV.

WiFi - It's all about the speed

When it comes to 'streaming', the most important component is your internet connection, specifically, the speed. Most people get their cable TV, phone and internet as a package deal. *(Bundle)* And most streaming devices use WiFi but, like my Roku Ultra, there are a few that also have a wired *(ethernet)* socket so you can simply plug in your network cable and be up and running from the get go.

That brings us to the next consideration, the actual speed you are getting from your cable provider. At a bare minimum, to cover most streaming services like Hulu, Amazon Prime, Sling, Netflix, or YouTube TV, you'll want to have a download speed of about 8 Megabits per streaming device. Meaning, if you want to stream to two TVs at the same time, you'll need twice that or 16 Megabits. However, for DIRECTV NOW, figure 12 Megabits per streaming device for a really solid experience. It's a good idea to check with other content providers to see what their recommendations happen to be, for instance, fuboTV currently recommends 20 Megabits per device. Higher quality streams like what's referred to as 'Ultra HD' will require even faster speeds, Netflix pegs this at about 25 Megabits per device.

To find out how fast your connection is, you can go to your cable provider's website and find their tool to perform a speed test. Otherwise, you can check the website you want to stream from, for instance, Netflix has a link to their speed test on their website. Or, you can go to a website like www.speakeasy.net to test your download speed. If your download speed is greater than you need, you're most likely good to go.

After you get your Roku all set up, it has a handy download speed indicator buried in the settings. After everything is all set up and done, you can find the instructions for locating the speed indicator on Page 75.

Let's assume your internet speed is up to snuff, the next consideration is the kind of streaming device you want to use. Like I mentioned before, this booklet will concentrate on using a Roku as the streaming device, because that's what I used to get myself up and running. But, you can use whatever your preference might be. Remember, the beauty of all of this is that there are a lot of different choices and you can change how you're doing things at any time because, with streaming, you're not tied down to any particular device or any particular service.

How to setup and use a Roku.

First off, now that you know what a Roku is, how do you get your hands on one? Well, you can get them directly from the Roku website. (www.Roku.com) Or, you can do what most everyone else does and get one from Amazon or even WalMart the next time you're strolling through the store. Point is, unlike a pocket T-shirt in a decent color, they're easy to get. *(Seriously, why are all the pocket T-shirts in my size either grey or dark green?)* As I mentioned before, there are a lot of Roku TVs being sold nowadays and since Roku is built into the TV you can skip past the rest of this 'How to' section and go straight to *'Now, back to setting up the Roku'* on Page 36.

Note: One of the really underplayed advantages of a Roku TV is the ability to add an external antenna to the TV to pick up any missing 'over the air' or broadcast local channels available in your area.

There are several models of Roku. *(Throughout all of this, other than learning that Roku means '6' in Japanese, I've never been able to find out what is the plural of Roku is. Rokus, Roki, Rokuses?)* In any case, most Roku follow the same general setup steps. You'll probably think most of this is pretty self explanatory, but because this book is aimed at first timers and people looking to make a decision, I'll go over it a little bit anyway.

Once you get your Roku out of the box, you'll need to connect it to your TV. If the Roku you settled on isn't one of the 'streaming stick' type, *(They tend to look like a pack of gum that was assimilated by the Borg.)* that plug directly into an HDMI port on your TV or, if the model you bought doesn't happen to come with one in the box, you'll need to get your hands on an HDMI cable to connect the Roku to the TV.

Which leads to a word about HDMI cables.

> *Okay, as much I wanted to avoid it, I have to get ever so slightly technical. If you don't have an older TV or you aren't having any issues plugging everything in and getting it going, you can skip this next section and jump right to 'Now, back to setting up the Roku' on Page 36.*
> *Seriously, you won't care...*

HDMI stands for **H**igh-**D**efinition **M**ultimedia Interface. It's the current standard way of connecting a device *(Like a Roku or a DVR, or a Cable Set Box.)* to a TV. Most HDTVs have more than one HDMI port on them to allow you to connect several different devices for convenience.

HDMI cables are used to connect your device to your TV. Because the signal it transmits is digital and not analog, an HDMI cable is either good or bad with no gray area in between like any of the other kinds of cables you may be used to. This means that paying a ton of money extra for a theoretically higher quality cable probably won't necessarily matter. But, that being said, the HDMI specifications have been incrementally upgraded over the years and a cable's particular capabilities aren't that clearly marked on the cable in most cases. The current HDMI specs are setting at Version 2.1, but most often the 2.0a version is what you'll see in the wild. I bring this up because generally, the higher the HDMI version number, the bigger, the faster the data can flow. So an older HDMI cable made for HDMI version 1.4 is the equivalent of sipping through a straw, while HDMI 2.1 would be like drinking out of a firehose. So the HDMI cable you use has to be of the same standard or higher to be able to handle the HDMI signal you need.

But, a newer streaming device is designed to communicate with reasonably new TVs and the HDMI part of the equation isn't necessarily the only culprit if your HDMI connection doesn't happen to work with an older TV for instance. That's because, the other part of the HDMI equation is HDCP, which stands for **H**igh-bandwidth **D**igital **C**ontent **P**rotection and is the mandated copy protection scheme to make sure you're not able to intercept and copy audio or video signals as they travel between your device and your TV. Just like HDMI, HDCP has different versions and this can be a problem if your new streaming device is being used with a much older TV that has an older HDCP version. The Roku is forced to follow the HDCP copy protection rules that are built into it and can stop the signal if it and the TV can't settle on a version of HDCP they can both agree on. If that's the case, you won't see anything on your TV no matter how good of a cable you have.

It used to be that, if your TV is too old, you'd need to get a newer TV or maybe an older, possibly used, streaming device, assuming that you could locate one to match the TV specs. But, as luck would have it, for an older TV that either doesn't have HDMI inputs or an older incompatible version, Roku actually has model with 'component' outputs. *(Those are the yellow, red and white RCA or phono jacks you see on the back of some TVs.)* The Roku built to solve this problem is called a **Roku Express+**. Its component outputs solve the connection issue and are more guaranteed to work with that old TV. There are companies that make HDMI to component adapter cables, but they generally only transmit the digital signal and since component *(RCA)* signals are usually analog, your TV still may not work with it. In that case, you would actually need an HD Video Converter Box to perform the conversion between the two cables. Generally speaking, acquiring the **Roku Express+** model with its component outputs is going to be cheaper and a lot less painful than buying and using a shelf full of adapters.

Now, back to setting up the Roku...

Once you get the Roku connected to the TV, make sure that the TV is set to display the HDMI input you're using. HDMI ports are labeled on your TV so you should be able to use the TV's remote control to change to the HDMI port you are using for the Roku. Most TV remotes will have a button labeled as 'Source' or 'Input' to allow you to change to the HDMI input port you need. Down the road, after the Roku is up and running there is an option that will allow it to automatically change the TV to the right HDMI port as soon as you press a button on the Roku remote to wake it up. *(You can find that information on Page 76.)*

Note: By design all of the Roku streaming devices are 'always on' devices. This allows them to check for and automatically keep themselves updated so they're always ready to use.

Next you'll need to place the batteries in the remote. I've set up several Rokus and, right out of the box, all the remotes automatically communicated with the Roku. But, if yours doesn't, it's a simple matter to use the 'pairing' button in the remote's battery compartment to get them talking. Push in the 'pairing' button until the light flashes and release. In about 30 seconds or less the Roku should be paired with its remote.

The Roku remote control is pretty self explanatory with the usual circle of directional arrows pointing ∧ **up**, ∨ **down**, < **left** and > **right**. These are mainly used to navigate through menus. The **OK** button, which works the same way that clicking on 'OK' works when you're on a computer, is either located at the center of the directional arrows on the remote or on older models, located just below them. The **OK** button is used to choose an item after using the arrow keys to navigate to its location on the screen.

Above the directional arrow keys near the top of the remote is another key with a left pointing arrow on it, this is the ← **Back** button and it can be used to step backward through a menu or eventually exit an 'app' *(Roku refers to these as channels instead of apps like everyone else.)* and take you backwards to the main Roku menu.

Which leads us to the ⌂ **Home** button, that would be the button at the top of the remote that has the little house symbol on it. The ⌂ **Home** button operates similar to the ← **Back** button, except it closes the channel *(App)* you're running and takes you back to the main Roku menu in one move.

Now that we have the basics out of the way, it's time to power up the Roku by plugging in the power adapter to the unit *(Streaming sticks can often get their power from a handy USB port adapter on the TV.)* and the indicator light on the Roku will begin to flash as it boots up. As long as you have the proper HDMI port chosen to display on your TV, within a few seconds you should see the Roku dancing letters followed by the main Roku screen waiting for you to press the **OK** key on the remote.

> Note: The Roku website mentioned earlier now has excellent videos to walk you through the initial setup process.

At that point the Roku should walk you through the internet connection process unless you have a wired connection. If it doesn't, you'll need to use the arrow keys to navigate to the **Settings** menu and choose **Network** to setup your Wireless (WiFi) network connection.

Once you're connected to your network the Roku may install the latest updates and restart itself. When it restarts, the first time setup wizard should take you through all the main things that are required, like setting up the remote to work with your TV and your preferred display settings. *(The Automatic display setting usually works best.)* The last step in the setup wizard is to link your new Roku to your Roku account online by going to the Roku website online via your computer and typing in a code that's being displayed on your TV screen. Going through the process of using the website link displayed on the TV will not only activate your Roku and link it to your account, it will walk you through the process of setting up a Roku account if you don't already have one. A Roku account is required and, once it's linked, you can even use it to manage your Roku through the Roku website and automatically purchase pay channels through the Roku store.

> *Make sure to only use the website link as it's displayed on the TV screen. Don't rely on a search using terms like 'activate Roku' because there are reports of bad actors out there that are launching websites that spoof the Roku website to intercept your account setup and use it to get you to pay them to activate your Roku.*
>
> *You should know that while a payment method is included in the real Roku setup for future purchases, no money is ever charged to activate your Roku.*

Before you leave the Roku website, be sure to add some channels to your Roku. In fact that's one of the great things that sets the Roku apart from the rest are the number of channels they have available. Installing them is really easy and there are a ton of them. In fact, you might want to go ahead and add some of the other streaming services just in case you want to check them out in the future, like Sling, fuboTV or YouTube TV. Once you add them here, they will automatically download to your Roku and appear on the Home screen. *(You can always add more directly from your Roku later as well, see Page 40.)*

When your Roku is all updated and ready, use the arrow keys on the remote to navigate to the Roku Channel and it will help you get the lay of the land to see how things work. Roku is in the process of expanding new choices with their Roku Channel so you will continue to see a lot more things showing up there in the future.

Remember, if you can't figure out how to go backwards in any area, try the ← **Back** button at the top of your remote to move back to a prior screen. If you just want to leave a channel entirely and get back to the Home screen, use the ⌂ **Home** button.

> *Note: There is reliable information out there that the Roku Channel's design will be the new look replacing the layout of the current Main Menu screen via a major update in the near future. Stay tuned....*

Adding DIRECTV NOW

I began my project while my original cable provider was still connected. Like I mentioned earlier, the primary reason was because, years before, I had already bought and connected the Roku to my TV setup. It was occasionally used for Netflix, Amazon Prime and a few other movie services. At the time I began using it, no one was really even talking about 'Live TV' streaming so that wasn't even a consideration.

However, because my Roku was already setup and working, it seemed like a natural way to add a 'Live TV' streaming service and see if it was worthwhile. I did a lot of research to find out which streaming service offered the particular channels we watched the most in a bundle of channels we happened to like and that's why I decided to take DIRECTV NOW up on its one week trial run just to see if it was up to snuff.

Here's where things get interesting. Once you actually start the trial run, you should do everything you can to stick to using it exclusively, as much as possible, and avoid using your old cable setup. This is especially important if you have a 'Remote Ninja' in the house who's used to flipping through all the channels instead of using the on-screen guide. The fact that channels take a certain amount of time to change from one to the next, will be their biggest complaint. (*My wife refers to this as 'Going back to the '70s'. We didn't have cable in the '70's but it's not really a good idea to point that out...*)

    The idea is to immerse yourself in the experience to figure out whether streaming is your cup of tea. If you backslide, and keep going back to your old cable, it'll make the decision too difficult.

Setting up the DIRECTV NOW service

DIRECTV NOW is able to be viewed across many different kinds of devices, including your computer, so since it's not a Roku exclusive experience, you'll need to go online and set up your DIRECTV NOW account first.

For instance, my first stop was to jump on my computer and go to www.directvnow.com. *(I imagine you can also do this on a phone or a tablet, but I'm a computer guy, remember?)* This was where I chose what channel lineup bundle I wanted to start with, based on the channels I wanted be able to see. Once I figured that part out, I signed up.

If you've ever bought or signed up for anything online before, you shouldn't have any problem with the sign up part. In addition to setting up a payment option, you will need to create a username and password for access. Make sure you keep track of the password information you create in this step because you will need it to sign into the DIRECTV NOW channel once you add it to the Roku.

> *Note: Be sure to mark the date you signed up with DIRECTV NOW on a calendar. If the 'Trial' doesn't work out and you want to try something else, you'll be able to cancel in time to avoid a charge on your credit card.*

Once you're signed up, you can start up the Roku on your TV and if you haven't already downloaded the DIRECTV NOW channel when you were signing up, navigate down the **Home** screen menu and choose **Streaming Channels** from the menu. This is where you can find the streaming channels that are available for installation on your Roku. Then you can either pick through the choices on the right hand side of the screen until you locate DIRECTV NOW or you can use the Search. Once you locate it, press **OK** on your remote and choose to **Add the channel**. It will download and the channel will be added to your **Home** screen choices.

Usually, when you add a channel it places the channel on your **Home** screen at the bottom of the list. If you're going to be using it frequently, you might want to move it up near the top of the list for convenience. To move a channel, simply highlight it and then push the ✱ **Options** button on your Roku remote. An on screen menu will appear. Highlight **Move Channel** and press **OK**. Now you can use the arrow keys on the remote to move the channel all around the Roku Home screen until you have it where you want it, then press **OK** again to lock it in place.

Now that we have the DIRECTV NOW channel on the Home screen, it's time to fire it up by clicking on it. First time up, you should be greeted by the sign in screen. This is where you'll need to input the same username and password you created when you created your DIRECTV NOW account online. Highlighting each field and pressing **OK** will bring up an 'on screen' keyboard that will allow you to enter the information. When you are finished entering, highlight **Done** and then press **OK**. Once you make it through the sign in process you'll be logged in and the channel will load. The DIRECTV NOW channel should always remember your information and keep you logged in so you won't have to sign in every time you start it up. However, if you sign out, or remove and replace the channel on your ⌂ **Home** screen, or it installs some sort of major update for the channel, it may prompt you to log back in again.

Getting Started With DIRECTV NOW Streaming

Once DirecTV Now loads, it will start playing one of the channels. Changing channels can be done several different ways. Unlike your traditional cable companies, DIRECTV NOW lays out its channels in alphabetical order. That means, as soon as a channel appears on your screen, using the > **right** or < **left** arrow keys on your remote will take you to the channels that are adjacent to the channel you're watching, with each button press moving, one channel at a time, either up or down the channel lineup. Like I said, channels are laid out in alphabetical order, A through Z. So pressing the < **left** arrow will change to the next channel up the list towards A and the > **right** arrow will take you to the next channel down the list towards Z.

Guide

While you are watching a program, pressing the ∧ **up** arrow button on your remote will pop up the program **Guide**. This will work anywhere the **Guide** is available, not just when you are already watching a program. If you encounter an empty screen with nothing on it, try the Λ up arrow on your remote to bring up the **Guide** again so you can choose a channel to watch.

Once you are in the **Guide**. You'll notice a thin blue vertical line running from top to bottom of the **Guide** that designates the current time. Use the arrow keys on the remote to move up and down the program listings and forward and backwards through the time listings. Highlighting and pressing **OK** on any program that has the thin blue vertical 'time' line running through it will change the channel to that program in progress. When you're in the **Guide**, if you use the < **left** arrow to highlight the name of a channel and press **OK,** you'll move to a separate channel guide with the times and listings of programs for just that channel.

To exit the **Guide** without changing the channel, use the ← **Back** button to return to the program you are watching.

Main Menu

Pressing the V **down** arrow on your remote will bring up the **Main Menu**. The default view is the **Watch Now** area and usually, below it is **What's on** now. The **What's on now** area will have thumbnail pictures of the currently playing programs it recommends based on the programs it believes you like. I've noticed over time that will eventually learn your preferences and make better recommendations. You use the can use the V **down** arrow on your remote to move the highlight down from **Watch Now** to focus on one of the program thumbnails and move left and right to bring other programs into view. You will also notice that one of the choices is **Guide** and if you highlight it and choose **OK** on your remote it will take you right to the **Guide**. Occasionally, **What's on now** may recommend a program that actually already played earlier in the day, so if you highlight it and press **OK** on your remote, the real currently playing program on that channel will load. So it's better to look for the channel in a thumbnail as opposed to a particular program to get back to the channel you want.

> *One of the most frequently mentioned items missing from the DIRECTV NOW channel is the ability to switch back to the 'last channel' with a single button on the remote. Their response is that you can use **What's on now** to flip back to the last channel.*

In the **Main Menu** the next choice to the right of **Watch Now** is **My Library**. This is for recordings and bookmarks and I will cover them later.

Just to the right of **My Library** is **Discover**. This is a section organized by type where you can browse around looking for interesting things to watch. Some things are immediately available 'On Demand', some are on extra pay channels, like HBO. Pay channels will prompt you to subscribe. Some are even future programs that you can schedule to record.

> *Which leads to another positive surrounding DirecTV Now as a service. AT&T currently owns HBO. Because of that, the DirecTV Now is currently the least expensive service when it comes to adding HBO to your regular bundle at an extra $5 a month. Also, since you can make that change a month at a time, you can add HBO to your lineup when your favorite show comes into season and drop it when it's done.*

To the right of **Discover** is another way to go to the **Guide** and to the right of that is the 🔍 **Search** magnifying glass. If you're having trouble finding something, highlight the search magnifying glass and press **OK** on your remote to enter the **Search** screen where you can use the on screen keyboard to type and see your results.

Farthest to the right of the choices in the **Main Menu** is the gear symbol for **Settings**. In here you can set a few **Preferences**, use **About** to find the installed version of the DIRECTV NOW channel that's running or go to **Help** to get the link to the DIRECTV NOW Support website and lastly **Log Out**.

Overlay Screen

Pressing **OK** or the ⏯ button on your remote while you are watching a channel will overlay a screen with choices that can be highlighted at the bottom of the screen. If you do nothing, the overlay will time out and go away after a few seconds or you can press the ∧ **up** arrow on your remote to leave the overlay immediately.

Normally, the overlay opens with the square ☐ **Pause** symbol highlighted in the bottom center of your screen where pressing **OK** will pause the program and change the symbol to an ▶. Pressing the **OK** or ⏯ button on your remote will resume the channel.

When you are in the overlay, you can use the > **right** arrow button on your remote to highlight the 💬 symbol at the lower far right of the screen where pressing **OK** will bring up the **Closed Captioning** choice on your screen. You can choose to either turn it on, or cancel. *(There is also another way to get to the **Closed Captioning** option that I will cover later, on Page 65.)*

If you use the > **right** arrow button on your remote you may also be offered the **REC** choice which can be used to record the program. Sometimes if the **REC** choice isn't available, you can change to another channel and change back again to make it operational again.

Next to the **REC** choice is a ↵ **Replay** arrow. If a program in the **Guide** has one of these arrows appearing to the right of the title in its grid box, that means you can highlight, press **OK** and use ↵ **Replay** in the **Overlay Screen** to restart that program from the its beginning and play it as an 'On Demand' program. Like all 'On Demand' programs, however, you usually cannot fast forward through any of the commercials.

Pressing the < **left** arrow button on your remote will bring you to the ⓘ or **Information** choice. Pressing **OK** will open the **Information** screen with a more detailed description of the program you are watching. I'll cover the other choices that appear on the **Information** screen shortly. To leave the **Information** screen, use the ⟵ **Back** button on your remote.

Channel View

If you are browsing for a program in the **Guide** and use the < **left** arrow button on your remote to move the highlight further left past the channel name, a **Channel View** menu will appear that will allow you to choose which channels you want to display in the **Guide**. But, before we dig into that, move the highlight all the way to the bottom of the list using the V **down** arrow on your remote to bring **Jump to Day** into view and press the **OK** button on your remote. This will produce a list of the days you can jump ahead in the **Guide** to see what's coming up by date.

Bear in mind, if you choose to go too far into the future using **Jump to Day**, it may take a while to download all the future **Guide** information and, in some cases, it may actually be beyond the scheduled information that's currently available so the boxes in the **Guide**'s grid view will remain empty.

Back to the **Channel View** menu. At the top of the menu is the **Channels** header with **All Subscribed** and **Favorites** tucked in underneath. **All Subscribed** is the default view and means just that, the **Guide** displays all the channels you are subscribed to in the channel package you purchased. **Favorites**, on the other hand, only displays the channels you designate as your Favorite channels. To create a list of Favorites, hit the ← **Back** button on your remote one time to step backwards to highlight an individual channel name *(Not a program.)* in the **Guide** and press **OK** to open an individual channel's guide. You'll notice that one of the choices in the left hand side of the screen is **Favorite** with a small heart next to it. Highlight it and press **OK** on your remote. The heart will light up and that channel is now one of your Favorites. Use the ← **Back** button to exit the individual channel guide. You can follow this process to create your list of **Favorite** channels and use the **Channel View** menu to display only your **Favorites** in the **Guide**.

Also, in the **Channel View** menu is the **On Now** header with **Everything** *(The default view.)* **Movies**, **TV Shows**, **Sports** and **Kids** underneath. Choosing any of these will change the **Guide** to display the type of programs you choose.

The Online DVR

The **Online DVR** is similar to, but works a bit differently than a DVR you may have used from your cable company. Primarily because the DVR from your cable company is storing the recorded program on a hard drive inside the DVR/Set Top Box. *(Sometimes referred to as a Media Server.)* This means that, in order to record a program and watch a program at the same time, the box needs to be able to tune in to two channels at the same time. To be able to do that it means it needs to have two or more tuners inside the box. The more channels you can record simultaneously, the more tuners it needs to have. But, once something is recorded, it stays on the DVR indefinitely, usually until you delete it or until you run low on storage where it may delete the oldest thing to free up space.

An online DVR, on the other hand, is essentially doing the same thing, only storing everything online but there are some some notable limitations. Meaning, the recordings are earmarked up in the DIRECTV NOW cloud and are simply played back and streamed down to you when you play them back. This is similar to watching programs 'On Demand' but the difference primarily seems to be that recorded programs always allow you to fast forward through commercials. Also, 'On Demand' programs on some channels aren't immediately available for playback after broadcast because the particular channel provider, like CBS for instance, wants you to opt for their individual pay service. Bear in mind there are even some 'On Demand' channels that normally play minus the commercials. There also seems to be a limitation involving playing back a program that's still in the process of recording. It can be done, but as of now, it doesn't automatically switch to the live program should you skip some commercials and catch up to the live stream.

      With traditional DVRs the provider controls the number of channels that can be simultaneously recorded by limiting the number of tuners built into the cable box. So far, I haven't seen a limitation to the number of programs I can record simultaneously with the Online DVR, but since there's only 20 hours of recording time DirecTV Now may be betting you'll run out of space pretty quickly so what's the problem with as many simultaneous recordings as you want.

> *DIRECTV NOW shows that the DRV function is in 'Beta'. Meaning that it may gain or lose functions while they are developing it. Currently, the default amount of recording time is 20 hours of recordings and a recording can only be stored for up to 30 days. You'll find this 30 day limitation is a recurring theme when it comes to playing certain available programs. (More on this later.)*
>
> *You can always choose to buy more storage space by kicking in an extra $5 a month.*

How to Record a show using the DVR

There is more than one way to record a show using the DVR. I'll go through them one at a time in the next sections.

Record a show from the Guide

Use the ∧ **up** arrow on your remote to bring up the **Guide.** Navigate to a future program. That means a program that isn't currently playing. Programs in the **Guide** that don't have the thin blue vertical 'time' line running through them are upcoming programs and you can mark them for recording by highlighting them and pressing **OK** on your remote to bring up their information screen. When it opens, one of the choices you will see is **Record**. When you highlight it and press **OK** on your remote a second box will pop up where you can click **OK** to confirm the recording or optional **Record the series**. Choose **OK** and it will take you back to the Information screen and pressing the ← **Back** button on your remote will exit the screen and return you to the **Guide**.

If you choose **Record the series**, another screen will open asking if you want to record **All Episodes** or **New Episodes**. Highlight your choice and move the highlight down to **Record the series**. Choose **OK** and it will take you back to the previous Information screen and pressing the ← **Back** button on your remote will exit the screen and return you to the **Guide**.

> *Note: If you choose to record a series, you should know that currently, the DVR Beta has some difficulty discerning between new, never before broadcast shows in a series and older reruns of that series that may be running in syndication on other channels. This can result in a lot of unwanted recordings of previous programs in a series and if you aren't careful, they can fill up your storage space. For non-syndicated shows, this shouldn't prove to be much of a problem, but if you notice it happening, you may need to cancel recording the series and set each episode to record one at a time.*

How to cancel the recording of a show or a series using the Guide

Essentially, canceling a recording from the Guide follows the same steps as above.
Use the ∧ **up** arrow on your remote to bring up the **Guide.** Use the arrow buttons to navigate to the program you previously set to record. Once you locate and highlight the program, press **OK** on your remote to bring up its Information screen. When it

opens one of the choices you will see is **Cancel Recording**. When you highlight it and press **OK** on your remote it will cancel the recording and it will switch the setting from **Cancel Recording** to **Record**.

Recording a program while it is in progress

While Pressing **OK** or the ▶❙❙ **play or pause** button on your remote while you are watching a channel will overlay the screen with choices that can be highlighted at the bottom of the screen. If you do nothing, the overlay will time out and go away after a few seconds or you can press the ∧ **up** arrow on your remote to leave the overlay immediately.

Normally, the overlay opens with the square ☐ **Pause** symbol highlighted in the bottom center of your screen where pressing **OK** will pause the program and change the symbol to an ▶. Pressing the **OK** or the ▶❙❙ **play or pause** button on your remote will resume the channel.

If you use the > **right** arrow button your remote you may also be offered the **REC** choice which can be used to record the program. Sometimes if the **REC** choice isn't available, sometimes due to the type of program that is, but most of the time you can change to another channel and change back to make it operational again. Highlight **REC** and press **OK** on your remote.

Depending how long the program has been playing will determine whether the program will be recorded from the beginning or if it will start recording at the time you actually started recording.

How to play recorded program

To play back a program that you have recorded. Press the V down arrow on your remote to bring up the **Main Menu**. In the **Main Menu** the next choice to the right of **Watch Now** is **My Library**. Use the > **right** arrow on your remote to move the highlight to **My Library** and press **OK** on your remote. This will display thumbnail pictures of some of the programs that you have recorded. Highlighting a thumbnail picture and pressing **OK** on your remote will begin playing the program or if there are other recordings of a series it may bring up an Information screen where you can choose which recorded episode you want to watch.

You can also choose **View All** which will place the recorded programs in list view where you can also choose which one to play. You can also navigate to the topmost choice on the list called **Upcoming**

Recordings and press **OK** on your remote. This will display all the upcoming scheduled recordings. If there is a program set to record that you would like to cancel, highlight the program and press **OK** on your remote to bring up the **Information** screen and choose **Cancel Recording**. To leave the **Information** screen, use the ← **Back** button on your remote.

Viewing a recorded program

Once you start playing a recorded program you can use the ▶❘❘ **play or pause** button both start and stop the playback. The ▶▶ **Fast Forward** and ◀◀ **Rewind** button on your remote function the same as you are used to with other devices and pushing either one up to 4 times increases their speed. The screen freezes during these operations so you can't see the progress, but you can see the time to the far right of the timeline to keep track of how much time is elapsing during the procedure. This makes fast forwarding through commercials almost more of an art than science but, after a while, you do catch on. I've even heard talk of making a drinking game out of it, ...make up your own rules.

At the end of playing back a program you are offered a choice at the bottom of the screen to **Delete** the program. Simply use the arrow buttons on your remote to highlight it and press **OK** on your remote. If the screen goes blank, simply press the ∧ **up** arrow on your remote to bring up the **Guide** and you can choose a program to watch.

Bookmarks

Pressing the V **down** arrow on your remote will bring up the **Main Menu**. In the **Main Menu** the next choice just to the right of **Watch Now** is **My Library**. Highlighting **My Library** and pressing the V **down** arrow twice on your remote will take you to the **Bookmarks** area. This is the area where you can manage and use things you have Bookmarked. You can Bookmark a program using its **Information screen** no matter how you get there. For instance, while you are watching a program, pressing the ∧ **up** arrow on your remote will bring up the **Guide** where you can browse for a program you'd like more information on. When you highlight the program and press **OK**, the **Information screen** is displayed with **Bookmark** as a choice where you can highlight **Add to Bookmarks** and press **OK** on your remote.

Bookmarks are similar to Favorites in a web browser on your computer. The main difference is that you can even **Bookmark** something that doesn't currently exist as a program in the **Guide**, but that you might be interested knowing about if it becomes available later.

You can set a **Bookmark** from any program's Information screen. For instance: use the ∧ **up** arrow on your remote to bring up the **Guide.** Navigate to a future program that you're interested in. That means a program that isn't currently playing. Programs in the **Guide** that don't have the thin blue vertical 'time' line running through them are upcoming programs. Highlight the program and press **OK** on your remote to bring up the program's Information screen. When it opens one of the choices you may see is **Add to bookmarks**. Highlight it and press **OK** on your remote to add it to your **Bookmarks**. However, some programs may not show the option on the **Information screen**. In that case, highlight **View all**

episodes and press **OK** on your remote. This will display a second screen with the ability to **Add series to bookmarks** as a choice. Highlight it and press **OK** on your remote. To leave the **Information screen**, use the ← **Back** button on your remote as many times as needed to get back to the **Guide** or your current program.

How to work with your Bookmarks

Pressing the V **down** arrow on your remote will bring up the **Main Menu**. In the **Main Menu** the next choose **My Library** by highlighting **My Library** and pressing the V **down** arrow twice on your remote to take you to the **Bookmarks** area. The programs you have added to your Bookmarks will appear as thumbnail pictures. To work with a bookmarked program, highlight a thumbnail picture and press **OK** on your remote. You can choose an episode to view, remove the bookmark, change to a different season, etc.

Most of the shows will be available 'On Demand'. The rest follow the rules of the channel provider and unfortunately, that means that not all of them are available for playback all the time.

The second way to turn Closed Captions off and on

This is actually a part of the Roku settings so it works in other areas as well. Push the ✱ **Options** button on your remote and the **Options** menu will pop in from the upper left hand corner of the screen. The uppermost item in the **Options** menu allows you to turn on **Closed Captioning** by using the > **right** or < **left** arrow keys on your remote to step through the choices of **On Always**, **On Replay** and **Off**. The second choice on the list is **Audio Track** which seems to be locked on **Default** on my setup so I'll skip that one. If you happen to have a Roku Ultra, the third choice in the **Options** menu is **Volume mode** and the choices are **Off**, **Leveling** and **Night**. *(Formerly referred to as Night Listening.)* Again, use the > **right** or < **left** arrow keys on your remote to step through the choices. You should take a little time to experiment with the settings to find the one you like. But, in our case I should note, **Night** changes the sound to allow you to hear voices better at lower volumes. *(Which is a particular favorite of Maverick...)*

The final choice in the **Options** menu is **Accessibility**. A > **right** arrow choice here opens a submenu with all the accessibility options tucked in underneath. You can even have the Roku talk to you with every button push you make, and using the voice and volume of your choice.

Miscellaneous Stuff

Here are a few things that aren't really covered by the other sections that do stand on their own.

What if something isn't working right?

When it comes to technology, any technology, the one thing you can count on is things not always working the way they are supposed to. DIRECTV NOW is a 'live' channel and should always be considered a work in progress. Which is as it should be so that they can implement any improvements to make it work better and better as time goes on. I can tell you that, in the time I've been using it, it has evolved from having 'major', to 'minor' annoyances and now we're at the 'occasional' annoyance level. But, the solution is usually always the same.

If something isn't working right, for instance, the **Guide** shows empty slots instead of program names, or an option in the **Overlay Screen** is faded out so you can't choose it, or if there's any other thing that doesn't seem to be working right. Simply, wait for the right occasion, *(Like a commercial.)* press the △

Home button on the remote to bounce back out to the Roku menu and restart the DIRECTV NOW channel. Once you're back in the channel, you should be back to normal.

But, let's say things are really, really just not working right. It's time to follow the first rule of Roku - 'When in doubt, reboot'. If it's convenient, you can simply unplug the power cord for the Roku either at the box or the wall outlet and plug it back in. Or, if you don't feel like getting up and walking across the room and, if you're in a channel, like DIRECTV NOW, press the ⌂ **Home** button to return to the main Roku menu and navigate to and highlight the **Settings** menu. Press the > **right** arrow button on your remote, highlight **System** and press the > **right** arrow button again and look for **System Restart**, highlight it and use the > **right** arrow button one last time to highlight **Restart** and press **OK**.

The Roku will reboot and the problem should be cleared out.

Take it with you

Okay, obviously I'm not talking about the great beyond. Everyone knows you can't take it with you, no matter what it is. But, terrestrially speaking, one of the things you can do with your Roku that doesn't get a lot of play is travel. When you get to your destination, as long as you have a reasonably good internet connection, you should be able to use it just like you do at home. Simply plug it in, make sure the TV is set to the right HDMI port and you're in business.

But, *(Rule number one with technical explanations: There's always a 'but'.)* if the location, *(Like a hotel, for instance.)* restricts streaming access, all bets are off. That's where you can always try the online DIRECTV NOW solution you can find on Page 80.

Headphones

The headphone jack on the side of the 'enhanced' remotes, like the one with the Roku Ultra, is definitely cool. Since the remote communicates with the Roku Ultra via RC *(Radio Control)* you can plug in a set of headphones and wonder around quite a distance away from the TV and still hear what's going on. But, there are a few things you need to know about.

> It should be noted that, nowadays, not all Roku models have the headphone jack on the remote. Currently the **Roku Ultra** does as well as some of the older models. But, don't despair if the headphones aren't an option on your Roku. There's a way you can have your cake and eat it too using the **Roku Mobile App** that works on Apple and Android phones turning them into super enhanced remotes plus a whole, lot more. I'll cover them more on Page 76.

If you accidentally happen to press the volume control on the side of the remote while you are trying to unplug the headphones, the normal volume controls will end up in limbo and you'll need to completely restart your Roku to get them going again in order to control the volume on the TV. Also, if possible, I find it's generally better to unplug the headphones when no sound is taking place, like when a channel is loading to help avoid it.

The batteries in the remote provide the power for the headphones when they are in use. As a result, if you're a heavy headphones user, be prepared to go through 'AA' batteries at a pretty good clip.

The single most optional accessory

The Roku Ultra and some older Roku models come with the ability to add a micro SD card. In fact, if you add a lot of channels from the Roku Store you may see a suggestion on the screen about adding a micro SD card to speed up loading times. But, the question becomes whether you should bother to or not when you could simply highlight any channels you're not using and use the ✻ **Options** button on your remote to pop up the Options menu and choose to remove the channels from your menu solving the problem.

But, if you have an older model, the answer to the 'add a micro SD' message is simpler, don't bother. Older Rokus are made for older types of micro SD cards that are getting harder to get your hands on and that makes the chances of getting a compatible one pretty slim.

However, if you have a newer Roku and you think you want to try it, don't waste your money on a large capacity, I know the micro SD card in my Ultra is only using 1.2 Gigabytes of space on average and word on the street is that they can't use much more than 2 gigabytes of space anyway so all the extra space on a high capacity micro SD card would simply be wasted.

The micro SD slot is located above or below the HDMI plug on the Roku so you'll need to power it off and unplug the HDMI cable to be able to access it. It can only fit one way with the copper 'fingers' pointing inward so if you don't feel the 'click' when you attempt to install it, flip it over and try it with the opposite side up.

Once it's installed, reassemble and power up your Roku. It may start up with a notice asking to format the micro SD card. Allow it to format and, when it completes, you should be able to load as many channels as you like on the **Home** screen without seeing the 'add a micro SD card' message any longer.

If it doesn't present you with a 'formating' message wait a while and reboot the Roku and it may detect it and prompt you to format the micro SD card. Also, don't be surprised if on occasion you're asked to format the card again. The cached data it keeps on the card will simply be redownloaded when the formatting operation finishes.

DIRECTV NOW Channel Timeout

If you don't have a reason to touch the remote for about 4 hours, *(Volume, Guide or Main menu, for instance.)* the DIRECTV NOW channel will time out, stop streaming the channel and ask if you are still watching or wish to exit the channel. This is normal and apparently has been instituted to stop the stream in case the user might have a data cap and is wasting bandwidth unnecessarily.

Setting up the Automatic HDMI port for the Roku

      The Roku can be set to automatically energize the HDMI port on the TV that it's connected to when you turn on the TV using the Roku remote. If the remote doesn't have the On/Off option, pressing any button after the TV is on will switch the HDMI port on the TV to the one connected to the Roku.

      To set the option, start at the **Home** screen menu and navigate to the **Settings** menu. Press the > **right** arrow button on your remote, highlight **System** and press the > **right** arrow button again and look for **Control other devices**, highlight that, press the > **right** arrow button, highlight **One Touch Play** and press **OK** to choose it. You can then press the < **left** arrow button on your remote repeatedly until you get back to the **Home** menu.

      While this option works on most newer TVs, some older TVs sometimes will require that you restart them in order to change back to a different HDMI port making this another one of those YMMV situations. *(Your Mileage May Vary.)*

The internet download speed indicator

The latest Roku operating system version includes a handy internet speed indicator you can use to see if your connection is performing well. From the main Roku menu, navigate to and highlight the **Settings** menu. Press the > **right** arrow button on your remote, highlight **Network** and press the > **right** arrow button again and look for **About** and your speed and the date it was last tested on will show next to **Internet Download Speed**.

Roku Mobile App

The Roku Mobile App is advertised as a way to turn your phone or tablet into a mobile version of your Roku remote. There are both Apple and Android versions available in either the Apple Store or Google Play Store. *(Whether it's Apple or Android, just search the store for 'Roku' and you'll turn them up.)* I'll do everything I can to avoid getting into too many specifics because both versions are getting updates on a fairly regular basis and that can cause changes to the way they act and appear. So I'll stick to the basics.

There are, however, a few things you need to know about the app in order to understand what to expect. For instance, before you install the app, you have to make sure your phone or tablet is connected to the same WiFi network your Roku is using because the remote control of the Roku is handled via WiFi. Because it's making a WiFi connection to your Roku, it can control the Roku but, at least for now, not your TV. That means that it can't turn your TV off and on or control the volume since those aren't handled via WiFi but rather IR *(infrared)* which makes them independent of the Roku. Since there are already a lot of other phone and tablet apps out there that can control TVs, I fully expect to see that capability added to the Roku Mobile App in the future.

After the app is installed on your device and you open it for the first time it will perform a search on your WiFi network for your Roku. If you have more than one, you'll need to pick the one you wish to work with. You can always switch to a different Roku from inside the app's 'Settings' area any time you want. But normally, if you have more than one Roku on your network you will be prompted to choose which one you want to work with every time you start the Roku Mobile App anyway.

Currently the app opens in **What's On** mode. So you get a screen with all the stuff that Roku thinks you'll be interested in. Like 'Popular Movies', 'New Releases', 'Free Movies', etc. These are live links, so any item you choose could change the channel on the Roku and take you to your choice. The reason it opens in **What's On** is because the different menu choices are lined up across the bottom of the app screen. Since **What's On** is the leftmost item, it's first up.

The second choice is **Channels**. Again if you touch **Channels** the Roku Mobile App will switch to an app view of the channels you see on the Roku's Main menu and the Roku will do the same on your TV.

Touching a channel on the app like DIRECTV NOW will start it running on your TV and the app will switch into the **Remote** mode.

Which brings me to the **Remote** mode choice that appears at the bottom of the Roku Mobile App to the right of channels mode which places it dead center of the mode choices at the bottom of the app. When it's in Remote mode the app screen switches to a virtual screen that mimics the layout of a Roku remote complete with an **OK** button and directional arrows. All the buttons on the remote work the same way they do on the real remote. In addition, they've added a couple of extra buttons. That's where the Roku Remote App shows its real stuff.

Let's say you and two or three of your closest friends need to watch TV together and you don't want to make any noise. It's easy enough for one of you to watch privately with the headphones on the remote of your Roku Ultra, but what about the rest of your crew? That's where the Roku Mobile App comes in. Install the app and set it up on up to 4 devices and you can all plug headphones into your phones or tablets and listen together by switching to **Remote** mode and touching the button that looks like a pair of headphones in the lower right corner of the app. Roku refers to this as 'Private Listening'.

There's also a **Photos+** mode that will allow you to cast photos and videos to the TV, set up your own personal Roku screensaver and send music to your TV. I won't get into these right now because it seems they are another one of those YMMV situations.

DIRECTV NOW online

One of the things you can also do with DIRECTV NOW is watch it on your computer. Luckily, most of the same things you learned with the Roku controls work the same way on the DIRECTV NOW website. The only differences is that instead of using your remote, you're using your mouse. Open your web browser on your computer and go to:

https://www.directvnow.com/watchnow

Log in with your user name and password and you'll be watching in no time. While it starts up running in a window, placing your mouse over the program you're watching will produce an overlay with all the screen controls to handle the **Recording**, **Closed Captioning**, **Volume**, *(Including Mute, which is currently M.I.A. on the Roku.)* as well as screen size. To go to a full screen you'll need to click on the expand icon ⛶ twice. Pressing the **ESC** key on your keyboard will exit the fullscreen view.

You'll also notice that the **Watch Now**, **My Library**, **Discover** and **Guide** choices are all there as well and work the same way.

> Tip: It is, by far, much easier and quicker, to manage your Online DVR recordings and **Upcoming Recordings** online.

When you are done viewing or managing things online, you can simply close your browser and go about your business.

But, more importantly, when you travel, let's say dragging around your Roku with you everywhere you go, isn't something you really want to do. DirecTV Online and a laptop will easily solve that problem. All you need to have is internet access and once you log in to DIRECTV NOW online, you're in business.

Now what?

It's up, it's working, you're pretty sure you can live with it, and now you've decided to keep streaming and kill off your old cable.

This leads to the, 'Now what?' question. Well, now it's time to turn off the cable TV portion of your cable bill. Since I don't know your particular set up with your cable company, I can't really tell you how your cable company will handle this. But, I can speak from my own experience and relate some things from others I've spoken to that have gone through it and how they handled it.

I'm talking about making the call. The one to your cable provider where you give them the bad news. *(Good for you, bad for them.)* You want to turn off your TV subscription but you need to keep your data *(Internet)* connection.

In my case, I contacted Frontier and told them my kids had finally moved out and my wife and I, since we weren't using the cable, decided we wanted to keep the internet and phone but drop the TV service. *(All of which was true, because we had already been streaming for a month.)*

So here's a bit of advice:

We're people. We're used to talking to other people on the phone to get things solved. But, when it comes to dealing with large companies and their farmed out support people with impenetrable accents from far flung corners of the globe, the answer is simple, - don't. If your situation allows it, try the 'Chat' function so many companies are including nowadays on their websites instead. Not because it's any quicker, but because it's flatout easier to state your case, understand what the person on the other end is really saying and actually get results. As a bonus, when you finish, they usually have some sort of method for you to keep a transcript of the whole conversation, just in case.

That's how I did it. I used the 'Chat with Support' link on their webpage to work with Frontier to handle dropping my TV service. Then I had a copy of everything that we discussed. If you're renting the cable set top boxes from your cable company they will probably want them back and since the rental fee is part of your cable bill you probably want them to have it back too. In the end, if you have a cable company as your internet provider, you may have to use their modem, but in a lot of cases you might even want to find out if you can purchase your own, especially if they are charging you for that as well. This is something you'll be able to tell by looking at your bill after it's all said and done. In my case, it took two billing cycles from Frontier to finally see what my normal bill for just data and phone would be.

Something to bear in mind is, it's the person on the other end of the line whose job it is to try to talk you out of cutting the cable even though all you're really doing is cutting your TV service. But, don't feel bad. The margins on TV service for the cable companies is really slim no matter what package they offer. The margins on phone and internet service, on the other hand, are where they're really making their money. By you cutting out your TV service, you are definitely not hurting their bottom line. At least until the time comes that you find a less expensive internet alternative.

I'll cover some future alternatives in a bit, but the next section is an additional area you may be able to save a few more dollars a month.

If you check your bill and find you are paying a monthly rental fee for a modem or a router and would like to save the additional monthly expense, there may be something you can do about it.

Unfortunately this brings us to another one of these:

Unavoidable explanation time

    If your cable provider connects to your house with a COAX cable and requires a cable modem in order to provide you with data access, you may be able to buy a modem and replace theirs to do away with the monthly rental fee on your bill. The cable company will not help you with the installation, or provide support for it, but may provide you with a list of modems they can recommend or the specifications they require for their network.

    If you have fiber, like FIOS, you may still have a COAX connection to connect to your router. Usually, this can easily be switched to an ethernet cable in what's referred to as the ONT (**O**ptical **N**etwork **T**erminal) which is a wall mounted box installed by your service provider which can be located indoors or outdoors where it translates the incoming fiber connection into a usable signal either COAX or ethernet. The ONT actually has a 'Customer Access' door that customer support may ask you to look at to help diagnose issues before they decide to send a technician to your house. This 'Customer Access' area is also where the ethernet port is located. If you can plug in an ethernet cable and run it to where your router is located, you can plug it into the 'WAN' port on your router, unscrew the COAX connection and be in business. In some areas, the ethernet port in the ONT isn't turned on. In that case, you can call and have customer support turn it on for you. Once it is turned on, you can ditch their router and buy one of your own. Since it's now an ethernet connection, almost any router will be compatible. Just make sure the WAN port on the router can handle at least the same speed as the internet speed you are paying for.

In either case, be sure to return the old equipment to the cable provider in order to have them stop charging you the monthly rental fee.

Looking to the future

Since there are more and more people shifting away from the cable and going to the streaming and over the air options, combined with more new streaming services coming on line all the time I think we're actually getting closer to the 'a la carte' streaming nirvana that all the cord cutters are longing for. Cable providers are starting to realize that they can actually join the fray by opting to go the streaming route for their services and drop the hardware. A few of the traditional cable companies 'channels' are showing up in the Roku Store and other devices. To top it off since mobile phones are becoming everyone's primary phone number, more people aren't necessarily sold on the need for a landline from their cable company any longer.

In a few years, even more change is coming because some of the providers are starting to roll out 5G and other super fast wireless technologies. These all have the potential to be game changers for streamers, possibly even lowering the cost of internet access through competition since a wireless data connection would bypass the need for cable or FTTH otherwise known as **Fiber To The** Home, like FIOS. Verizon has already kicked off their in home 5G wireless access service in some areas with others sure to follow.

Meanwhile, one of the other things in the mix are the LEOs. In this case, it doesn't stand for Law Enforcement Officers. It stands for Low Earth Orbit. There are already a few major players with this, for instance, there's OneWeb who will partner with HughesNet and then there's SpaceX which is Elon Musk's baby.

Traditionally, a communication satellite orbits the earth at a distance that is far enough out that it matches the speed of the earth's rotation. This makes it appear to us as though its position and never changes, when in reality it's at an altitude of 22,236 miles and traveling at a speed of just shy of 2 miles per second to remain stationary. That distance means that a signal from earth to the satellite and back takes quite a while to make the round trip. This long delay makes them less than desirable for internet access but, great for something like TV which is primarily a one way streaming signal.

This is where the LEOs are going to step in. The idea is to have low orbiting satellites and lots of them, meaning several thousand of them. The first batch will orbit at altitudes between 684 and 823 miles up. Since they're so close to the earth, the time it takes for a signal to make the round trip is almost negligible and the data speeds are much higher. Later, they're planning on satellites that will orbit even lower at 211 miles. SpaceX is so confident, they're projecting 500 megabit speeds and 40 million subscribers by 2025.

The really good news here is, even if you're not jumping ship to sign up with them, the fact that they're going to be providing an internet access service that essentially covers everyone, everywhere, will create serious competition with the providers you're dealing with now for internet access and that, in turn, should eventually help drive the costs down.

Epilogue

Okay, by now you realize that unless you plan on getting really creative, there aren't necessarily 50 ways to leave your cable company. But, as you can see, there are way more than 50 'reasons' to leave them. At least the cable part of it. No matter which way you decide to go, there are savings to be had. Especially once you add in all the taxes and fees, and ways they use to suck you in an sock it to you when your 'promotional period' comes to a crashing end.

Because of that, one of things you might want to do is keep a copy of an old cable bill around to remind you how much you're saving. And if you ever feel antsy to sign up with one of them again, drag it out and remember why you cut that cable to begin with.

www.ingramcontent.com/pod-product-compliance
Lightning Source LLC
Chambersburg PA
CBHW020601220526
45463CB00006B/2407